THE QUEEN'S UNIVERSITY OF BELFAST

A Giant's Stride in Engineering

A. JENNINGS

Professor of Civil Engineering

An Inaugural Lecture
delivered before the Queen's University of Belfast
on 8 November 1978

NEW LECTURE SERIES No. 115
ISBN 0 85389 156 7

Printed by
MAYNE, BOYD & SON, LTD, BELFAST

A GIANT'S STRIDE IN ENGINEERING

Introduction

Vice Chancellor, Dean, ladies and gentlemen, it is a quirk of University procedure that readers are appointed on the basis of what they have written. Similarly, if they ever have the honour of being appointed to a chair, they are required in an inaugural lecture to stand in front of the public to profess their discipline. Anomalous as these might seem, the tradition of the inaugural lecture is a very valuable one. I have often benefited as a member of the audience and now appreciate that it gives a new professor an excuse to step back from his current activity and try to view his own discipline in perspective.

In being appointed to this chair I will strive hard, and I will need to strive hard, to emulate the standard of achievement set up by my predecessor Dr Wells. He has had for a long time a worldwide reputation in welding and fracture mechanics and was energetic in pursuing many novel ideas while at Queens. These included a scheme for laying a seabed tunnel across the North Channel, a composite beam construction for easy erection of floors, a novel design for the internal structure of super-tankers, and a scheme for utilizing wave energy. This last investigation is being vigorously continued in the department under the direction of Professor Long. Since Ireland is surrounded by so much of this form of energy, its use could substantially reduce our demands on irreplaceable fossil fuels. Dr Wells has now returned to the Welding Institute as director-general and it is fitting that before leaving here he had the honour of being elected a fellow of the Royal Society.

Contrast with Aircraft Structures

Having had experience in the aircraft industry I would like to start by noting some outstanding differences between aeronautical and civil engineering. Since an aircraft needs to get off the ground the chief objective of the aircraft structural designer is to minimise weight. In contrast the designer of civil structures is only concerned about weight insofar as it affects cost. For this reason aircraft structures are mainly of aluminium and have intricately proportional members, while civil engineering structures are mainly of steel and concrete with members which are often more crudely constructed. Aircraft structures are designed by large teams who can afford to spend considerable effort because the aircraft will be mass produced. In contrast civil engineering structures are generally designed on a one-off basis by small teams or individuals who cannot afford to pursue such a protracted design process. A complete airframe is tested to destruction before any particular aircraft is registered as airworthy. In contrast the civil engineering designer tests structural components but is of course restricted to non-destructive tests on the complete structure. Hence he has less knowledge of the ultimate strength of his structures. For these reasons civil engineering structural designs tend to be more conservative, the engineer is less willing to risk new materials, methods of construction or design features. He is less willing to use computers to assist in analysis and relies considerably on simple formulae and on codes of practice.

Whereas aircraft structures have developed rapidly over a period of seventy years, with a considerable amount of stimulus from military requirements, civil engineering structures have developed more slowly over a period of more than 2000 years with most stimulus coming from commercial interests. However there are instances where spectacular progress has been made.

Early Development of Civil Engineering Structures

Prior to the industrial revolution the prime building material was stone, certainly of those structures which now remain for us to see (Fig. 1). Furthermore they were constructed, as far as we know, without

Fig. 1 The Pont-du-Gard near Nimes, a roman aqueduct bridge.

Fig. 2 The nave of Amiens cathedral.

strength tests on the materials and without any knowledge of structural analysis as we know it today. From the Egyptian pyramids and Roman arches through to the medieval cathedrals all construction was by trial and error. When we admire some of those Gothic cathedrals (Fig. 2) we would do well to remember that the flying buttresses are there to prevent the walls from being pushed outwards by the spreading action of the roof, and that pinnacles have often been included not as a pure adornment but to provide weight to stabilise the structure when subject to wind loading.[1] Cathedral collapses did occur. Men were gaining experience in the sense that Oscar Wilde used the word when he said 'Experience – the name men give to their mistakes'.

During the 19th century, revolutionary progress was made in structures. Iron and steel provided materials which were strong, homogeneous, durable, workable, and which could take large loads in tension as well as compression. These properties were vital to James Watt, enabling him to build pressure boilers and George Stevenson, enabling him to build strong and durable rails. They also gave much greater scope to the designers of bridges than did stone, which was only suitable for use in compression. This period of rapid progress can be exemplified by considering the medieval London bridge (Fig. 3) which had a drawbridge and nineteen irregular arches with spans ranging from 5 m to 11 m and which was the only crossing of the Thames at London till 1817. The waterway was so restricted that at certain states of the tide there

Fig. 3 London bridge 1209–1831, illustrated *c.* 1750.

was a $1\frac{1}{2}$ m. drop in water level due to the tidal race through the bridge. It has been described as a 'pierced dam' and according to tradition 'wise men went over, only fools went under'.[2] Yet in 1801 Telford put forward a proposal for a cast iron arch to cross the Thames with a single main span of 183 m (Fig. 4). Although this was not built, he did build the Menai suspension bridge with a clear span of 177 m which was opened in 1828, three years before the medieval London bridge was replaced (Figs. 5 & 6).

Fig. 4 Telford's proposed cast iron bridge to cross the Thames.

Fig. 5 Telford's Menai suspension bridge.

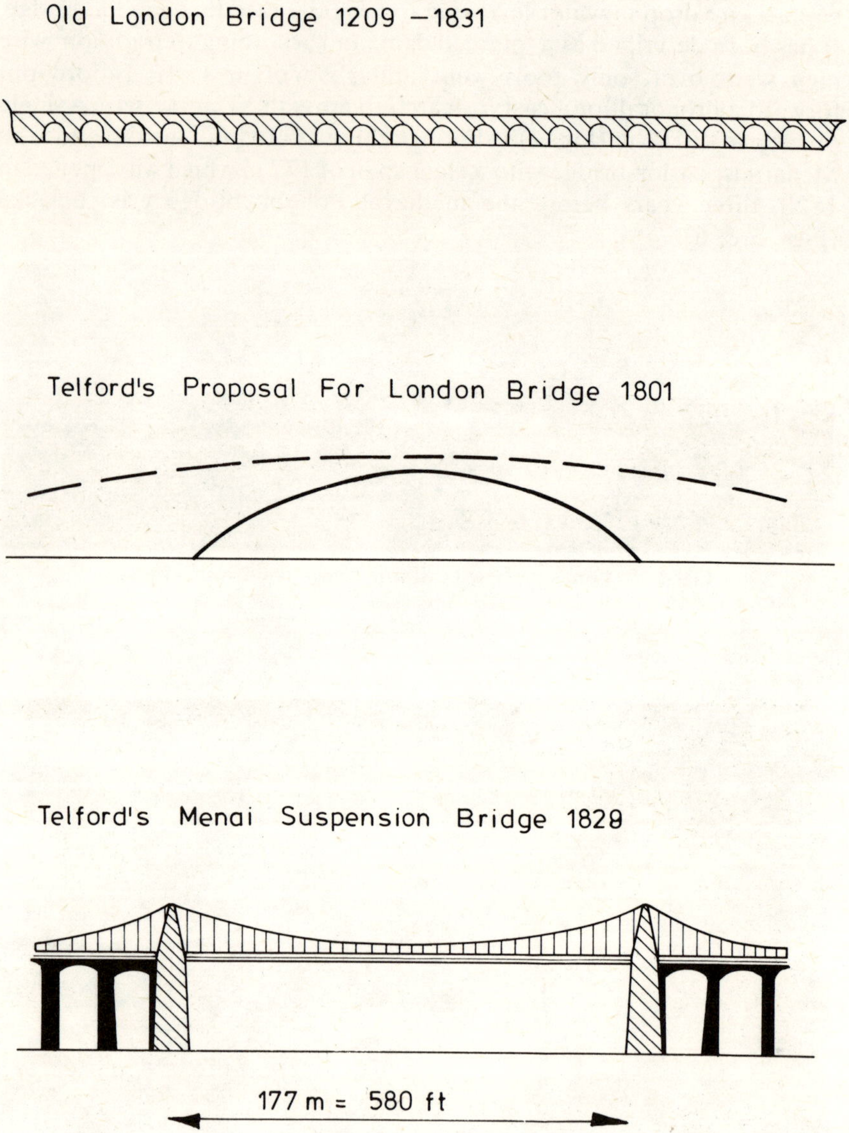

Fig. 6 Changeover from masonry to iron.

The Forth Railway Bridge

In the subsequent years of rapid development one of the most remarkable achievements was the construction of the Forth railway bridge which some American authors[2] have described as 'A giant's stride in engineering'. The Tay and Forth bridges were planned to shorten the East coast railroad from London to Aberdeen, so that trains could compete with those on the alternative midland route. The following background items are relevant:—

(*a*) Sir John Fowler and Benjamin Baker undertook the design after the disastrous failure of the first Tay bridge in 1879, which was attributed to a gross underestimation of wind forces. Benjamin Baker (who was later knighted) was responsible for the design of the bridge structure itself.

(*b*) Not counting suspension bridges, which were ruled out for rail traffic, the largest span up to that date was Eads' chrome steel arch bridge over the Mississippi at St Louis having a span of 159 m (Fig. 7). It may be noted that, although Eads' bridge finished up as an arch it was constructed using the cantilever principle.

Fig. 7 Ead's bridge over the Mississippi.

(*c*) There was no previous experience of bridge building with steel in Britain because a method of extracting the deleterious phosphorus, present in all local iron ores, had only just been invented.

Fig. 8 Forth rail bridge under construction.

Fig. 9 Forth rail bridge.

Fig. 10 A graphic demonstration of structural principles used in the design of the Forth bridge.

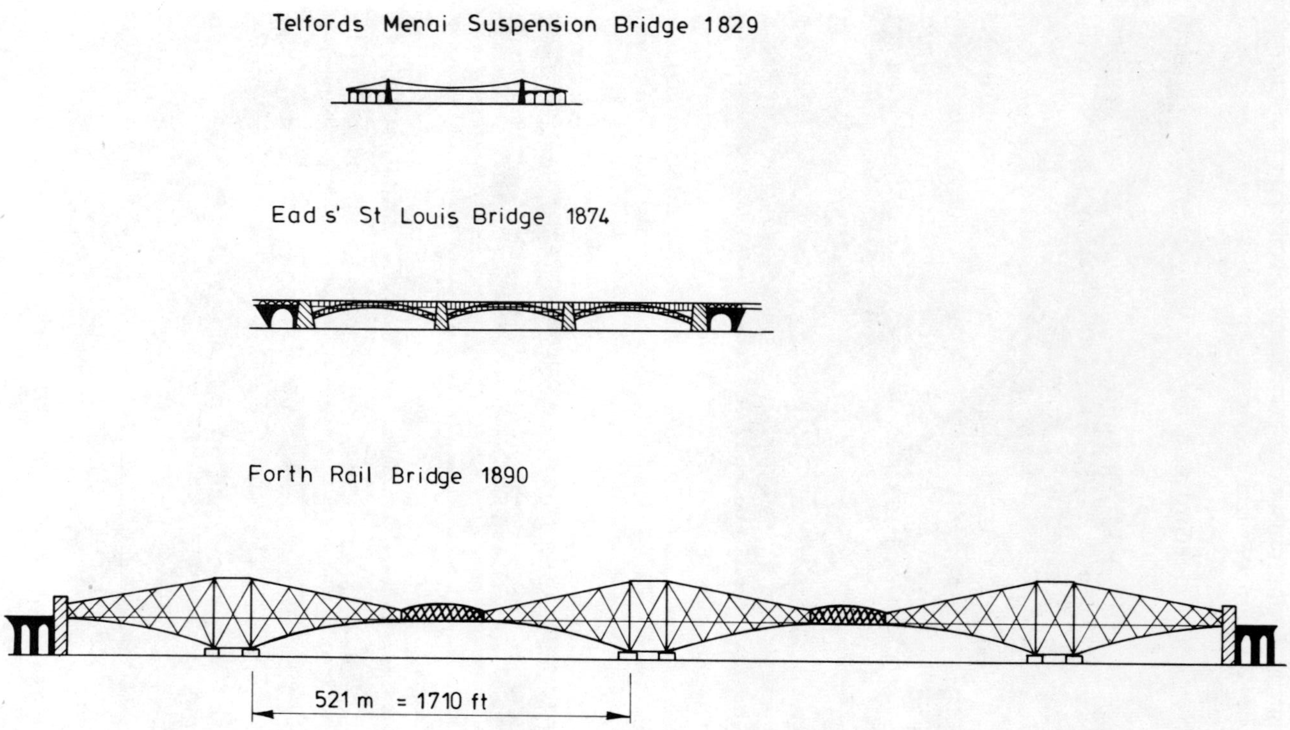

Fig. 11 The Forth rail bridge and contemporaries.

(d) Graphical methods of analysis were well developed for predicting forces in members of certain types of trusses and also overall deflections.

Although Baker is on record as saying that 'If an engineer has not had some failures with retaining walls, it is merely evidence that his practice has not been sufficiently extensive, for the attempt to guard against every contingency in all instances would lead to ruinous and unjustifiable extravagance', he certainly took as few risks as possible with the Forth bridge. Wind pressures of 2·7 kN/m² (56 lb/sq. ft) of projected area were used, being confirmed from site wind pressure measurements to be more than sufficient (this was about five times the wind strength allowed for in the ill-fated Tay bridge and corresponds approximately to present day practice). This gave lateral wind forces of 2000 tons on each of the main spans.[3]

Cantilever spans were used with tubular compression members and lattice tension members in order to keep down the areas subject to wind pressure. The legs of the cantilevers were splayed outwards in a 'Henry VIII straddle' to provide greater stability against wind load. The proportions of the members were all designed according to the available theory of structures. The three-dimensional geometry of the bridge must have made the design of joints in the tubular members particularly difficult. The resulting bridge, opened in 1890, had two clear spans each of 521 m, which is over three times the clear span of Eads' Mississippi bridge and to this day is only exceeded in length of span by the Quebec bridge, suspension bridges excluded. It is the only large span bridge over which trains are allowed to travel at full speed (Figs. 8, 9, 10 & 11).

Baker avoided any ornamentation so as to keep the whole work in harmony, and achieved an appearance showing strength, stability and, to some eyes, grace. But those who don't understand its design concepts may not claim it to be graceful at all. For instance, at the time of its construction William Morris said 'There never would be an architecture of iron, every improvement in machinery being uglier and uglier, until they reach the supremest specimen of all ugliness – the Forth bridge'.

Structural Forms for Large Span Bridges

In contrast to cantilever trusses, suspension bridges have always been generally acclaimed as elegant structures and are often presented as leading showpieces of engineering achievement. Structurally speaking a suspension span is just an inverted arch span with the vital difference that the members are being pulled in tension rather than pushed in

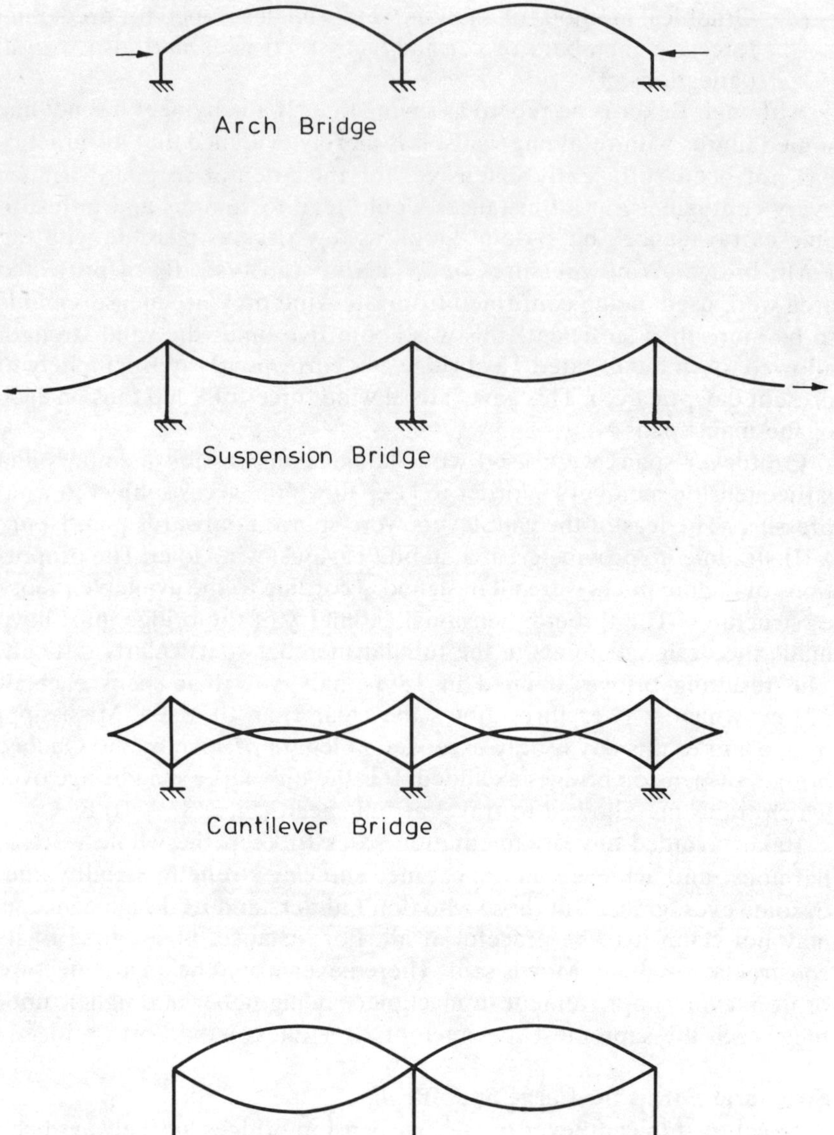

Fig. 12 Structural forms for large span bridges.

compression. Other large span bridges using the beam or cantilever principle (for example Brunel's Saltash bridge or the Forth rail bridge) are effectively combinations of these two forms (Fig. 12). Structures in compression need both strength against breaking and stiffness against buckling, whereas structures in tension only require strength against breaking to withstand static loads. An example of a structure failing in compression because of lack of stiffness is a slender tree trunk which bends over, so shedding its load without actually breaking (Fig. 13). If the ground could be inverted so that the tree hung down it would then be loaded in tension and that type of failure would not occur (Fig. 14). Although it is theoretically possible, without increasing its weight, to design any member so that it does not buckle in compression just by changing its cross-sectional shape (for example from a round bar to a tube or a double-walled tube as illustrated in Fig. 15), this tends to be impracticable for members which are very long. Hence the suspension bridge, which is in tension except for the towers, is the most effective of these structures for large spans. A further important advantage is that temporary supports are not required during erection as is the case with arch and beam type bridges. The fact that all of the world's 20 largest span bridges are of suspension type indicates their present superiority.[4] However this achievement has been painstakingly built up from a history of lack of understanding and spectacular failure.

Fig. 13 A buckled tree trunk.

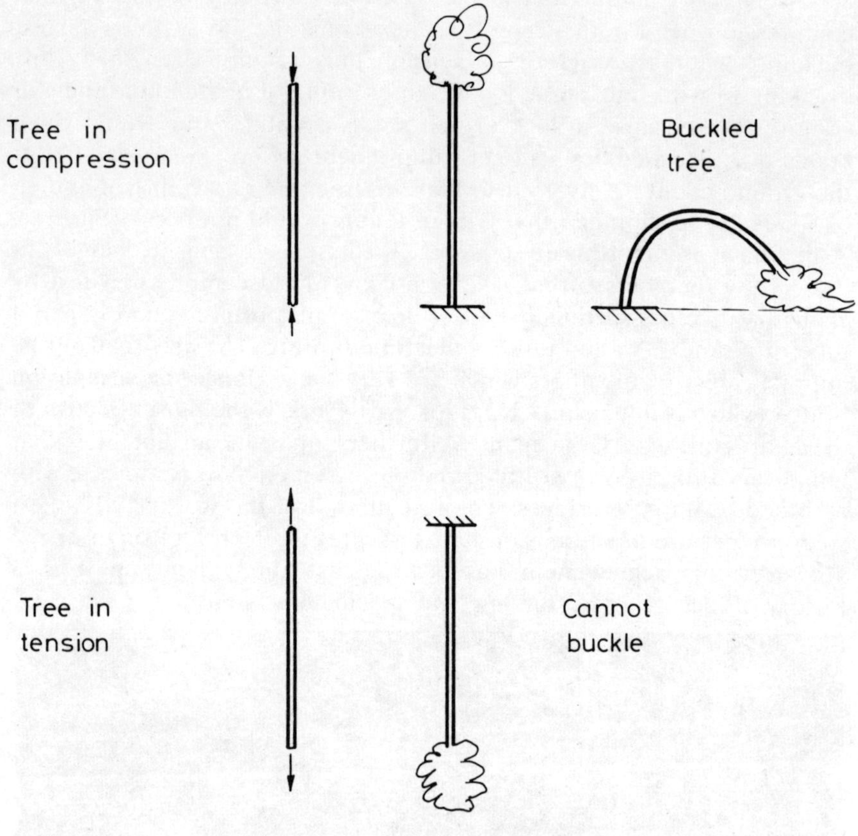

Fig. 14 Buckling as a compression phenomenon.

Fig. 15 Strength of columns of equal weight.

History of Suspension Bridges

The first person to put a level floor in a suspension bridge and hence make it suitable for traffic was an American James Finlay in 1801. During the 19th century many suspension bridges were built in America and Europe, some of which remain in use to this day, ranging from small bridges, such as the one at Ballievey near Banbridge (built by James Dredge in 1845) to the Brooklyn bridge at New York with a span almost equal to that of the Forth rail bridge (Figs. 16 & 17). However an alarming number suffered damage or catastrophic failure or were unsuitable for the purpose for which they were built, most problems arising because of their extreme flexibility, (or lack of stiffness) in the presence of dynamic loads.

Whereas a suspension bridge could be designed to carry its self-weight by simple analysis procedures based on the triangle of forces, even the statical behaviour due to traffic or wind load is difficult to determine and no methods of analysis were available in the early days. The dynamical

Fig. 16 Ballievey bridge, Co. Down.

Fig. 17 Brooklyn bridge, New York.

behaviour of such bridges, being more complicated still, could therefore not be rationally considered in the design process. A bridge at Newcastle-upon-Tyne intended for rail traffic had to be dismantled because of massive distortions of the deck as trains crossed it. Several bridges, one near Manchester, fell due to the oscillations caused by soldiers marching in step, and many bridges were severely damaged in winds. For instance Telford's Menai suspension bridge had to be strengthened after being damaged by wind soon after erection. Towards the end of the 19th century engineers were becoming very wary of suspension bridges and few were built.

In 1888 Melan developed the so called 'deflection theory' for the analysis of suspension bridges with vertical hangers and a deck girder under the action of any static traffic load. This form of construction (with vertical hangers and deck stiffening truss or girder) became the standard for many long span bridges built in America between 1900 and 1950 including the Golden Gate bridge at San Francisco (Fig. 18). These bridges were adequately designed to carry all loads applied statically and, since dynamic effects due to traffic movement are much less significant than for small span bridges, the engineers thought that their designs were reliable. However several of these bridges suffered vibration due to wind and in 1940 the newly completed Tacoma

Fig. 18 Golden Gate bridge, San Francisco.

Fig. 19. Wind excited vibration of the first Tacoma Narrows bridge.

Narrows bridge near Seattle which gained the nickname 'Galloping Gertie', lost its deck in a violent oscillation occurring at the modest wind speed of 40 m.p.h. (Figs. 19 & 20). In retrospect the American civil engineers should have learnt from their aeronautical colleagues who had by that time mastered the allied dynamic vibration problem called 'flutter' which affected aircraft since the earliest days of flight. Following

Fig. 20 Failure of the first Tacoma Narrows bridge.

the Tacoma failure the dynamical characteristics of suspension bridges were thoroughly investigated and now it is known how to design bridges so that they remain stable in even the most severe winds.

No major suspension bridges were built in Britain during this century until the construction of the Forth road bridge in 1964 which followed the style of the American bridges.

Modern Developments in Suspension Bridges

The British consulting firm of Freeman Fox and Partners have improved the design of modern suspension bridges by two features introduced into the Severn bridge and subsequently used for the Lil-lebelt bridge in Denmark, the Bosporus bridge and also the Humber Bridge which is now under construction. One feature is the use of a streamlined box girder instead of a truss to provide deck stiffness. This enables lighter deck and tower structures to be used and also gives benefits with regard to ease of erection and appearance. Secondly, since a welded box girder has very little inherent damping, inclined hangers

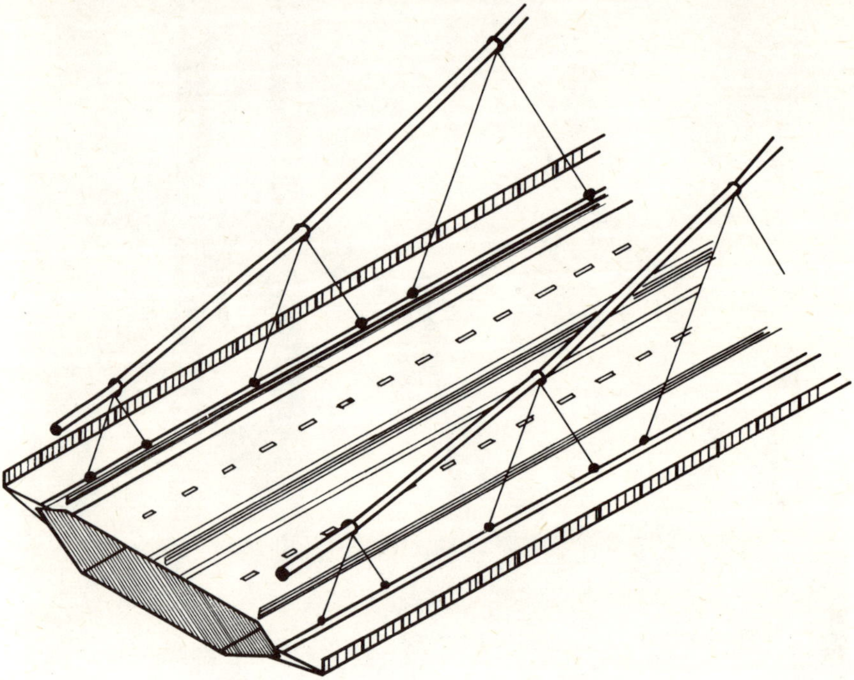

Fig. 21 Severn bridge deck.

have been adopted so as to increase the overall damping of the bridge (Fig. 21). It was only by using wind tunnels to analyse aerodynamic characteristics and computers to analyse the structural characteristics of the inclined hanger configurations that these two innovations could have

Fig. 22 A tower of the Humber bridge under construction.

been confidently put into practice. Once the Humber bridge is completed it will have a span of 1410 m which will be 121 m greater than the present largest span, the Verrazano-Narrows bridge at New York (Figs. 22 & 23).

Fig. 23 Catwalk on the Humber bridge prior to cable spinning.

However this is by no means the limit of possible achievement as far as suspension bridges are concerned. Freeman Fox & Partners are at present working on the design of a bridge to cross the Messina Straits between Italy and Sicily with a span of 3300 m (just less than 2 miles) which will make even the Humber bridge seem small (Figs. 24 & 25). In order to allow for wind load and also to guard against dynamic instability in wind with such a large span a completely novel type of deck structure is to be used in which a railway is to be carried at a lower level below the road deck.

Fig. 24 An artist's impression of the proposed Messina Straits bridge.

There is a theoretical limit for suspension bridges of about 6000 m which is based on the maximum span over which a cable will just support its own weight and which depends only on the strength/weight ratio of the cable. Economic considerations would preclude close approach to this limit. Table 1 shows the way in which the cable weight/deck weight ratio increases with span, from which it may be deduced that the Messina proposal will require massive cables and must be getting close to the practical limit. Indeed it is interesting to compare the Messina proposal which could advance the suspension bridge from $\frac{1}{4}$ to $\frac{1}{2}$ of the limiting span with that of the Forth rail bridge which in its day advanced the steel truss bridge from $^1/_6$ to $^1/_2$ of its limiting span. If successfully implemented the Messina bridge could surely be described as another giant's stride in engineering (Fig. 26).

TABLE 1 Cable weight penalty invoked by increasing the span of suspension bridges

	Span	$\frac{\text{cable weight}}{\text{deck weight}}$
Severn	986 m	0·28
Humber	1410 m	0·5
Messina	3300 m	1·4

Fig. 25 Model of deck structure for Messina Straits bridge.

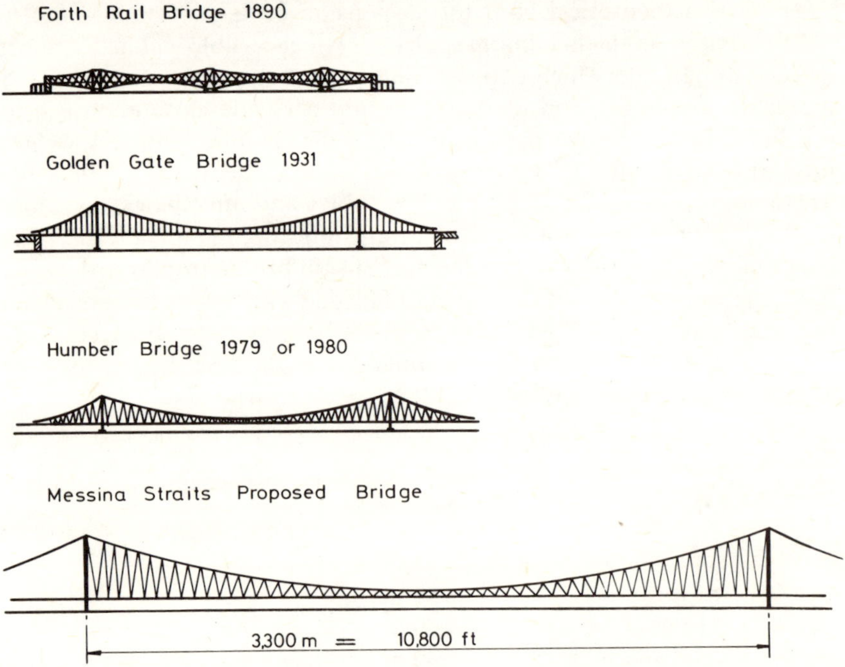

Fig. 26 The Messina Straits proposed bridge and its contemporaries.

Civil Engineering Department Work on Suspension Bridges

Here at Queen's University Professor Charlton initiated a research investigation on the structural characteristics of suspension bridges in 1963 and when I came in 1965 I carried forward from Manchester an interest in the subject. Since then model experiments and computer simulations have been carried out to examine the effects of different patterns of inclined hangers and more recently to examine Leonhardt's proposal for a monocable bridge (Fig. 27). Investigations have shown that to make effective use of the monocable concept it would be necessary to increase the cross-sectional area of the hangers considerably. But if this were done, it would enable the deck weight to be reduced by an even larger amount. However as yet no monocable bridge has ever been built.

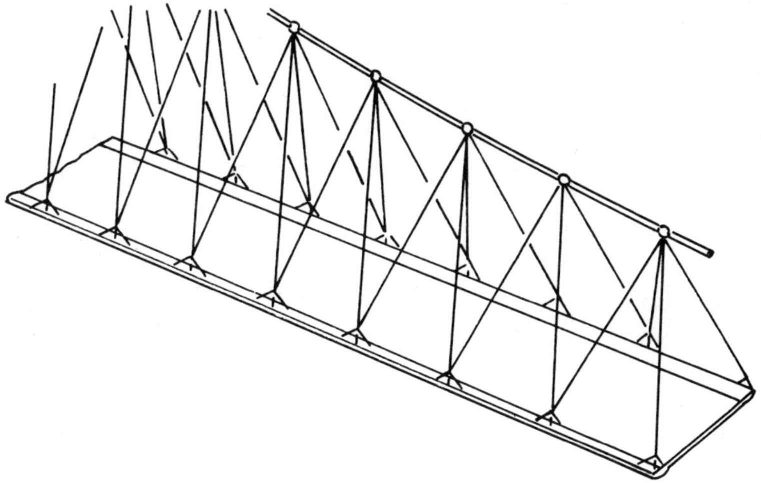

Fig. 27 Leonhardt's 1953 proposal for a monocable suspension bridge design.

At present the department is just commencing to undertake a research project in conjunction with Freeman Fox & Partners to investigate the structural characteristics of multicable bridges (Fig. 28). These should be stiffer than corresponding standard configurations and may also enable faster erection times, since parts of the deck can be lifted before cable spinning has been completed.

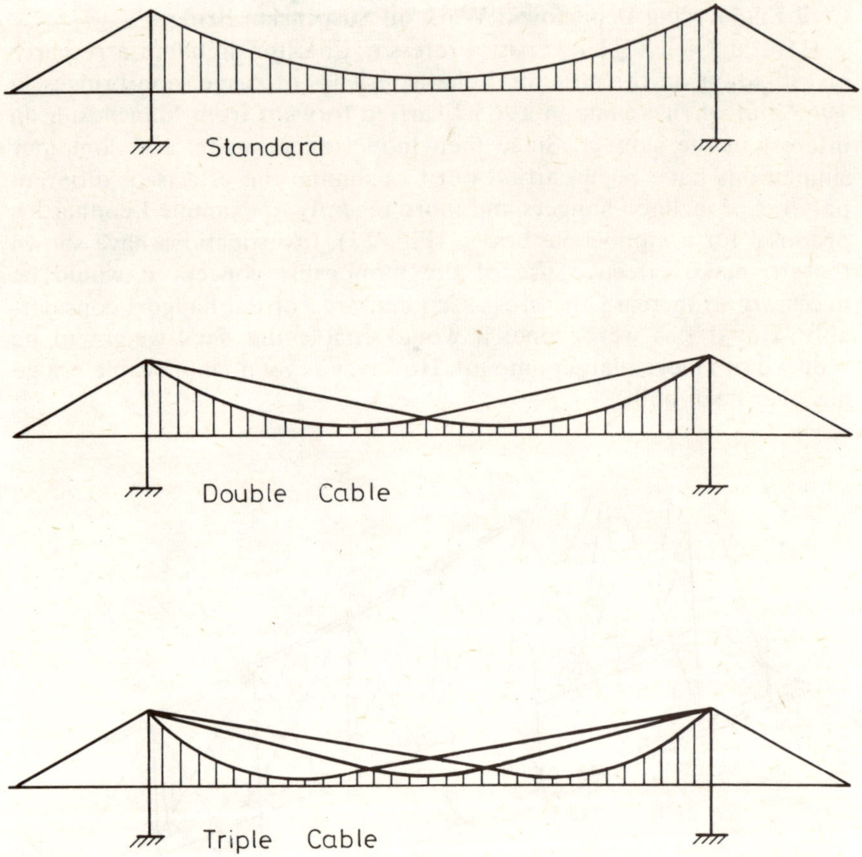

Fig. 28 Different suspension bridge configurations.

Other Developments in Bridge Structures

Whereas suspension bridges have been setting the pace for years by leaping larger and larger spans, they consist of but a small proportion of the bridges built throughout the world. Other types of bridge structures have come a long way since the days of the masonry arch. There are some elegant examples of modern large span arch bridges in both reinforced concrete and steel (Fig. 29). Even continuous beams can be designed for considerable spans when they are built of modern box girder construction (Fig. 30).

Fig. 29 Coldsprings arch bridge, California.

Fig. 30 Guanabara Bay bridge, Rio de Janeiro, of continuous steel box girder construction having a maximum span of 300 m. (Principal girder spans designed by Howard Needles Tammen and Bergendoff.)

Although some cable-stayed bridges were built in the early 19th century, there were some failures and the French engineer Navier expressed a preference for suspension bridges. No more were then built for over a century till 1955. However, since then they have achieved a spectacular revival and have been particularly useful in bridging the Rhine where spans of up to 350 m have been used (Fig. 31). It has been claimed that they could be used for considerably larger spans, and I am sure that this claim is well justified.

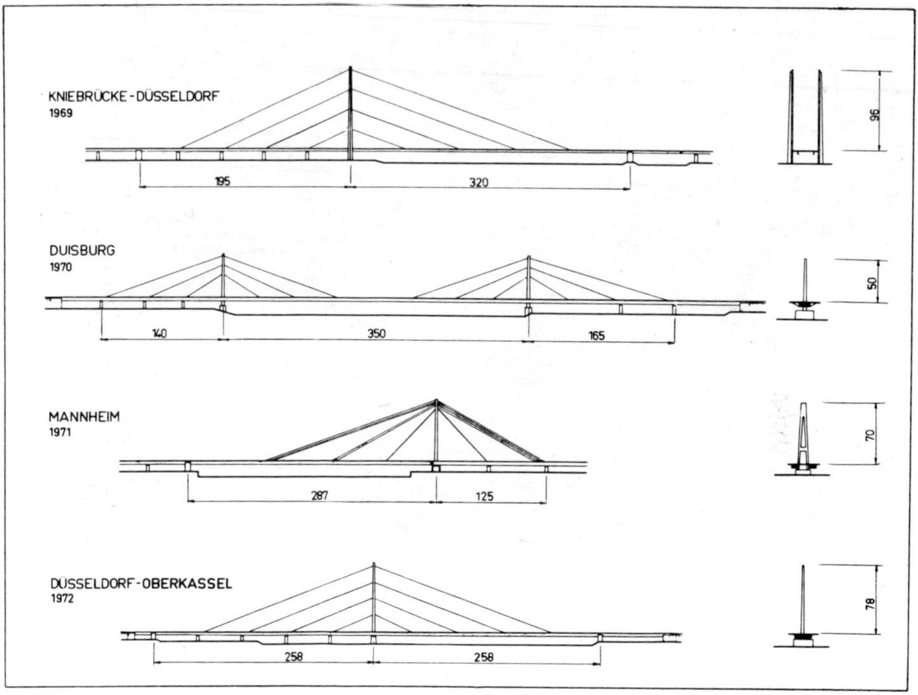

Fig. 31 Some cable-stayed bridges over the Rhine.

The Process of Bridge Design

A bridge engineer has many aspects to consider in the design process. Table 2 shows, in a broad way, the main parameters within his control and also the criteria which he uses to make decisions.

TABLE 2 Bridge design

Parameters	*Criteria*
Materials	Safety
Overall shape	Cost effectiveness
Member sizes	Ease of erection
Foundations	Ease of maintenance
Joints	Environmental effects
Method of Construction	Appearance

Fig. 32 Forth rail bridge detail.

Whereas the overall shape is what is most evident to us as we look at a bridge, the foundations, the design of joints and the method of construction are all important design parameters and may influence the overall shape in the long run.

Large structures can only be constructed by joining thousands of small parts together. The large tubular compression members of the Forth rail bridge were fabricated by boiler-makers who were expert at riveting curved plates (Fig. 32). Each main cable of a suspension bridge contains thousands of individual wire strands, the spinning of which is a highly skilled operation (in the case of the Humber bridge there are 14,948 strands in each cable) (Fig. 33). Concrete bridges contain a multitude of reinforcing bars and often prestressing bars, Shaw's bridge on the Belfast outer ring road being no exception (Figs. 34 & 35). Hence even bridges which look elegantly simple require extensive design effort and constructional skill to ensure that the various parts perform their correct function.

Fig. 33 Cable spinning at the Humber bridge.

Fig. 34 Reinforcing wires and prestressing ducts during the construction of Shaw's bridge.

Fig. 35 Shaw's bridge completed.

Simultaneous Equations

In the process of bridge design, and indeed the design of all large structures the engineer wants to know how much load is taken by each part of the structure – the technical term is 'stress'. He has to ensure that no part of the structure is overstressed in such a way that failure of it will cause, through chain reaction, a collapse of the whole structure. The accurate stress analysis of most structures, with the exception of certain types of truss (such as that used for the Forth rail bridge) can only be obtained by solving many simultaneous equations. Although 'classical' methods of solving simultaneous equations were available from the time of Gauss and had been used in astronomy and surveying, these methods would require months or years of hand calculation to be used for any but the simpler types of structural calculations.

All kinds of problems may involve the solution of simultaneous equations and to illustrate this for the benefit of those who are not familiar with structural mechanics or even O Level school mathematics consider the following simple conundrum.

Three years ago Jack was twice as old as Jill
but now he is two years older than she is.
How old is Jill?

A method of solution is to let x be Jill's age and y be Jack's age, then the information given may be expressed as

$$y-3=2(x-3)$$

and $$y=x+2$$

which may be converted to the two simultaneous equations

$$2x-y=3$$

and $$x-y=-2$$

From which the solution $x=5$ (which is Jill's age) can be deduced by subtraction.

In this set of two equations the solution was obtained with very little effort. However, as the number of variables increases, the difficulty of performing the last step in the solution process increases dramatically. Working without digital aids it would take about the length of this lecture to solve the set of four simultaneous equations

$$\begin{aligned} 8{\cdot}3x_1 - 3{\cdot}2x_2 + 0{\cdot}4x_3 - 2{\cdot}7x_4 &= 15{\cdot}33 \\ -3{\cdot}2x_1 + 13{\cdot}7x_2 - 5{\cdot}6x_3 + 0{\cdot}2x_4 &= 22{\cdot}49 \\ 0{\cdot}4x_1 - 5{\cdot}6x_2 + 10{\cdot}5x_3 - 2{\cdot}7x_4 &= -28{\cdot}53 \\ -2{\cdot}7x_1 + 0{\cdot}2x_2 - 2{\cdot}7x_3 + 9{\cdot}2x_4 &= 8{\cdot}12 \end{aligned}$$

and about one man year to solve 40 simultaneous equations by hand if the equations have no properties which allow you to take shortcuts. This is what restricted the use of the most effective structural design procedures. Indeed in the 1930's Hardy Cross, an American civil engineer, published the moment distribution method for solving structural equations which was more convenient and more economical to use than the 'classical' solution techniques.[5] This was welcomed by structural engineers who have used the method extensively ever since. Hardy Cross's method was the forerunner of the oddly named 'relaxation method' which has been used for many other applications other than structural engineering.

The mathematicians, on the other hand, extended the concept of matrices, first proposed by Caley[6] in 1857. If we consider the set of two simultaneous equations

$$2x-y=\ \ 3$$
$$x-y=-2$$

defining Jack and Jill's ages then in matrix notation they become

$$\begin{bmatrix} 2 & -1 \\ 1 & -1 \end{bmatrix}\begin{bmatrix} x \\ y \end{bmatrix} = \begin{bmatrix} 3 \\ -2 \end{bmatrix}$$

which can be expressed algebraically as

$$Ax=b$$

where $A=\begin{bmatrix} 2 & -1 \\ 1 & -1 \end{bmatrix}$ is the coefficient matrix

$x=\begin{bmatrix} x \\ y \end{bmatrix}$ is a vector of unknown ages

$b=\begin{bmatrix} 3 \\ -2 \end{bmatrix}$ is a vector of right-hand sides

The solution of these equations can then be expressed in matrix form as

$$x=A^{-1}b$$

where A^{-1} is called the inverse of A and can be computed from A according to set rules.

In the example $A^{-1}=\begin{bmatrix} 1 & -1 \\ 1 & -2 \end{bmatrix}$

and hence $x=A^{-1}b=\begin{bmatrix} 1 & -1 \\ 1 & -2 \end{bmatrix}\begin{bmatrix} 3 \\ -2 \end{bmatrix}=\begin{bmatrix} 5 \\ 7 \end{bmatrix}$ $\begin{matrix}\text{Jill's age} \\ \text{Jack's age}\end{matrix}$

Solving a set of simultaneous equations in this way may be considered as applying computational methods to matrix algebra. It results in a form of solution which is more cumbersome than classical solution methods and of no use to engineers who wished to solve large sets of simultaneous equations. As a result up till 1950, there was very little common ground or contact between such engineers and mathematicians who were interested in algebraic concepts.

Impact of Computers

As soon as digital computers became available they were put to use on the solution of structural equations. Structural frames for the Belfast Municipal Power Station West were among the first to be analysed by computer in 1953 (Fig. 36). The method adopted was developed by Dr Livesley[7] at Manchester for whom I had the privilege of working as a research student. The method he used would not have been suitable for hand computation but it was very successful on the computer since it enabled structural frames of any arbitrary geometrical form to be analysed with a single relatively simple program. In extending Livesley's method for frames to the solution of other structural problems, structural engineers developed the finite element method which is now widely used outside as well as inside structural engineering.

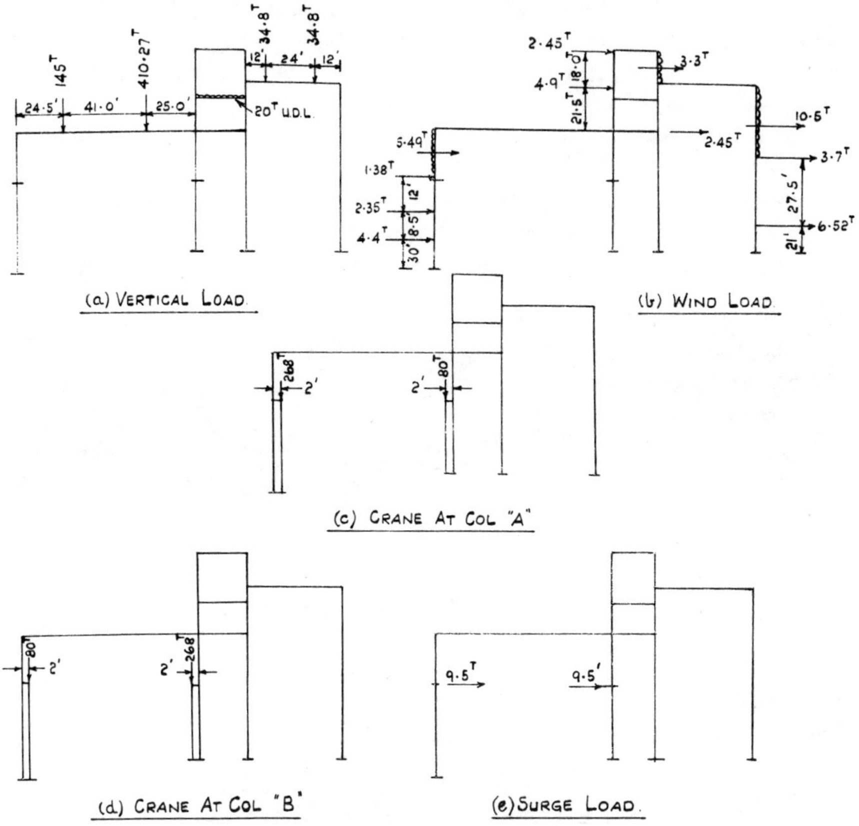

Fig. 36 Belfast Municipal Power Station West – 1953 frame analysis by computer.

When using these methods the resulting simultaneous equations have a sparse coefficient matrix, (in other words one with a large proportion of zero coefficients). In order to make effective use of computers it was therefore necessary to determine how to gain most benefit from this property in the solution process. As an illustration of the benefit gained through taking advantage of sparsity in the equations, an analysis of the space frame structure which now supports the cafeteria roof of the New University of Ulster required the solution of 730 simultaneous equations (Fig. 37). Sparse equation solving procedures reduced the storage requirement to 1/10th and the computing time to about 1/100th of what would have been required if conventional solution methods had been used. In terms of a computer charge, at the time, this involved a cost of £30 per analysis instead of something in the thousands of pounds. It is possible to obtain computer solutions to structural problems involving many more than 730 equations, however this example is an interesting one because the structural characteristics of this roof truss are not at all straightforward and cannot be predicted very well at all by other methods. Sparse matrix procedures developed primarily for use in structural engineering have also been used for analysing surveying networks, water supply networks, electrical supply networks, etc. Eigenvalue procedures developed for solving problems in structural vibration have also been used for principal component analysis and the

Fig. 37 Space frame roof at the New University of Ulster.

analysis of Markov chains – both valuable methods to the statistician. Hence structural engineers are not only making headway with computational techniques for their own problems, but are also adding considerably to the general pool of useful procedures.

Numerical Analysis

Since 1950 more mathematicians have become interested in numerical computation and as a result the effectively new discipline of numerical analysis has emerged. American numerical analysts, in particular, have made some important advances in methods of solving simultaneous equations and related eigenvalue problems, which are of value to structural engineers concerned with computation.[8] However, one of the problems of rapid technical development is that it induces a high degree of specialization. Numerical analysis is no exception as far as this is concerned. Most papers published in numerical analysis today are not the type of paper which non-specialists would understand. Furthermore there seems to be less contact than there used to be between numerical analysts and users of numerical methods (in engineering at any rate). Hence there is a real danger of progress being stultified through lack of communication. Professor Fox of Oxford University should be supported when he tells fellow numerical analysts that "Unless we are dealing with real problems, numerical analysis has no claim to life – it is not and should not try to be an abstract subject".

Here at Queen's University, numerical analysis is one of the topics covered in the M.Sc. course in computer science and applications. There have been some rewarding M.Sc. projects carried out in numerical aspects of structural engineering problems in recent years as cooperative ventures between the departments of computer science and civil engineering.

Use of Matrices

In the modern development of computational methods matrix notation has been invaluable. Since 1950, the number of published papers in structural computation alone which use matrix notation must be well into the thousands. Matrices were lifted out of the mathematician's back room virtually overnight and are now taught to engineers and even school children. However, their rise to importance has, in a way, been too rapid, because the teaching has been aimed at algebraic aspects rather than at its main area of application – computational methods. In some respects a little learning of matrix algebra can be a dangerous thing, simply because people tend to assume that once they know matrix algebra they can then, without much forethought successfully perform computations using matrices. Furthermore, since the intermediate steps in any digital computer computation are not inspected by the operator, there is a strong likelihood that longwinded or unreliable methods, once programmed will remain in use because their weaknesses are unlikely to be diagnosed during execution.

For instance it has been too often assumed that matrix inverses need to be computed. Hence there is a chapter on matrix inversion in most books on matrix computation and also there are inversion routines on all large computers. Yet, all this is unnecessary, since in situations where an inverse arises in the formal algebraic solution to a problem, it is always more efficient, and, where sparse matrices are involved, it is radically more efficient to use simultaneous equation solution techniques instead.[9] It took almost 100 years for people to learn the lesson that one should not apply computational methods to matrix algebra but instead apply matrix algebra to computational methods. Until matrix computation is taught along with matrix algebra some of the old mistakes will continue to recur.

Position of Mathematics and Computation

It is of vital importance for the engineer that he has good command of the mathematical and computational techniques which may be of particular use to him. With increasing accessibility of computers in the future, this is going to be of continuing importance. But in learning these techniques the engineer should not get involved with mathematical ideas and concepts which are of esoteric rather than practical value. Professor Gordon[10] of Reading University summed the position up well when he said that 'Mathematics is to the scientist and engineer a tool, to the professional mathematician a religion, but to the ordinary person a stumbling block'.

Mathematics deals with abstract ideas in an exact way, whereas engineering has to deal with the properties of real materials and the world surrounding them about which there is always some degree of uncertainty. Whereas improved computational methods may definitely be of use to him, the results of such computations can never be any more than a good guide to decision making.

Load Factors

Of the criteria used to assess bridge design, and indeed all structural design, safety is the most important. If the structure collapses it is no use worrying about the other criteria. Designing a structure is like sitting an examination in which the examiner tears up the whole script if he finds a single fundamental mistake in reasoning (This by the way is not a practice we resort to in the Civil Engineering department, although sometimes we may feel like it). Because of the importance of guarding against collapse, designers of structures have always left some margin between the level at which they expect collapse to occur and the maximum load they anticipate could act on the structure during its lifetime. This safety margin is generally referred to as 'load factor'. If the engineer has complete confidence over the design process a load factor of about 1·01 would be needed, giving about a 1% margin between the maximum possible load and the strength of the structure. However, there are assumptions in the analysis which are not completely valid, there is variability in material strengths and imposed loads, and it is impossible to erect any structure to precisely the required geometrical form. It is for these reasons that the structural engineer has to choose load factors greater than 1·01 (normally between 1·5 and 3), so that he is making most structures unnecessarily over strength in an attempt to guarantee that none should collapse. The load factor has the nickname 'factor of ignorance' and indeed it does reflect those aspects of the design process over which the engineer has incomplete control. Consequently the choice of load factor is to some extent a guess.

Seriousness of failure

One of the many important functions of codes of practice in civil engineering is to propose load factors, so that the engineer does not need to spend hours in philosophical debate before he initiates a design. The choice of load factors is therefore delegated to the code drafting committees who attempt to weigh up all evidence regarding the reliability or otherwise of all the design factors. Greater load factors are proposed where materials are more variable in strength (thus concrete is given a higher load factor than steel) and when design loads are more likely to be reached. However an important factor which could influence the load factor and which was proposed by Sir Alfred Pugsley[11] is the seriousness of failure. Pugsley suggests that if a structural collapse would involve large danger to personnel and/or would incur extensive economic costs then the load factor should be higher than would otherwise be the case and vice versa. Although Pugsley's proposal has not met with general approval, it should be given serious thought by engineers and maybe also by the public who have to pay the bills for excessively strong structures or suffer the consequences of failures.

Codes of Practice

In an environment where safety is important the codes of practice take on the role of specifying minimum standards to achieve a safe design and so there is a tendency for engineers to assume that a design to code specifications must be safe. Therefore codes need to be increasingly comprehensive. The new code CP110 on the structural use of concrete published in 1972 has the important clause.

'In addition, due to the nature of a particular occupancy or use of a structure (e.g. flour mill, chemical plant etc.) it may be necessary in the design concept or a design reappraisal to consider the effect of a particular hazard and to ensure that in the event of an accident, there is an acceptable probability of the structure remaining after the event even if in a damaged condition.'

Deciphering the legalistic language we could interpret this as 'Where accidental excessive loads may occur which go beyond normal design criteria then damage should be as localised as possible.

This clause, which was not in the previous code, covers the situation at Ronan Point in 1968 in which a gas explosion caused much more severe damage than it should have done to a block of flats, and would also cover, for instance, the case of a ship crashing into a bridge pier, as happened in Tasmania more recently.

However as codes of practice become more comprehensive they also become more complex. Increasing complexity is not welcomed by designers who have to spend more time in the design of a particular structure. Furthermore, since the committee drawing up a code of practice cannot be expected to envisage all possible future uses of the code, the onus in the end must still be with the engineer to consider the relevance or otherwise of the code. Codes of practice also make the job of the university lecturer more difficult because they appear to divorce design practice from structural theory and behaviour. Also much experimental research on structural behaviour may be of only academic interest unless it influences the drafting of new codes of practice. There seems to be a good case for expecting drafting committees to go much further than they do at present to indicate what information they took into account and what lines of reasoning they adopted in drawing up their codes. Whilst recognising that more elaborate codes must in some cases be justified, this move would help to restore the various links between design, theory and experimental research, and would encourage healthy criticism of codes where they are based on fallible or outdated reasoning.

Scope of Civil Engineering

I have been mainly talking about bridge design which is just one aspect of structural engineering, which again is just one aspect of the discipline of civil engineering. Since I came to Queen's University in 1965, over 850 students have graduated in civil engineering. In so many ways which we take for granted, the work of civil engineers impinges on our lives, not only when we cross bridges, but also for example when we sit in lecture theatres or auditoriums, when we turn on the tap and obtain relatively pure water, when we switch on the television and receive signals from a tall television mast or when we take our boat down to the marina and place it in the shelter of a breakwater. Civil engineers interact with people from many other disciplines. I have already mentioned links with mathematicians, numerical analysts and computer scientists but they also need to work with geologists in designing foundations and siting dams, etc, with architects in the design of buildings, with mechanical and electrical engineers in the construction of power stations and with chemists in the purification of water and development of building materials, to mention a few.

Every student who takes up a career in civil engineering hopes to obtain reward by knowing that he has helped to produce something of lasting benefit to his fellow man. All civil engineering projects have some element which makes them different from previous experience. Not every venture is a giant's stride in engineering, but it is at least one small step.

Epilogue

I would like to finish with a quotation about Benjamin Baker which gives some insight into the profession of engineering and helps to summarise some of the points I have been making. It has been said of him[3] and I quote "that his greatest service to his profession was the way in which he bridged the gulf between theory and practice by showing how both might work together if properly used. His great dictum was that all theoretical calculations involved assumptions that were often convenient rather than true; that they could be valuable as a check but disastrous if accepted with blind faith. Technology, he held, was therefore of little value unless it was accompanied by practical experience, sound judgement and bold initiative". Ladies and gentlemen, the Forth rail bridge stands as a constant reminder of this truth (Fig. 38).

Fig. 38 The Forth rail bridge.

ACKNOWLEDGEMENTS

The author would like to thank the following for permission to publish photographs and other illustrations: The French Government Tourist Office, London for Fig. 1, Scientific American for Fig. 2, the British Museum for Figs. 3 and 5, the British Library for Fig. 4, British Rail for Fig. 8, Dr P. T. F. Feenan for Fig. 16, the University of Washington libraries for Figs. 19 and 20, Messrs Freeman Fox and Partners for Figs. 22, 23, 24 and 25, Messrs Howard Needles, Tammen and Bergendoff for Fig. 30 and Professor Leonhardt, Stuttgart University for Fig. 31.

REFERENCES

1. R. Mark, 'The structural analysis of Gothic cathedrals'. *Scientific American,* Vol. 227 (5), pp. 90–99, Nov. 1972.
2. R. S. Kirby, S. Withington, A. B. Burr and F. G. Kilgour, '*Engineering in History,* McGraw-Hill, New York, 1956.
3. L. T. C. Rolt, '*Great Engineers*', G. Bell & Sons, London, 1962.
4. J. Virola. 'The world's greatest bridges', *Transactions ASCE,* Vol. 134, pp. 756–760 (1969).
5. H. Cross, 'Analysis of continuous frames by distributing fixed-end moments', *Transactions ASCE,* Vol. 96, pp. 1–156 (1932).
6. A Cayley, 'A memoir on the theory of matrices', *London Phil. Trans.* Vol. 148, pp. 17–37 (1857).
7. R. K. Livesley, 'Analysis of rigid frames by an electronic computer', *Engineering,* Vol. 176, p. 230 (1953).
8. A. Jennings, 'Matrices ancient and modern', *IMA Bulletin,* Vol. 13, pp. 117–123 (1977).
9. A. Jennings, '*Matrix computation for engineers and scientists*', Wiley, London, 1977.
10. J. E. Gordon, '*Structures – or why things don't fall down*', Penguin Books, Harmondsworth, 1978.
11. Sir A. Pugsley, '*The Safety of Structures*', Arnold, London, 1966.